Journal

Name:.......................................

Birthday:.................................

Address:...................................

Phone:......................................

Email:.......................................

SCAN ME

Canela Journals

Made in the USA
Monee, IL
28 May 2022

97151174R00090